I0810804

The INSIDE GUIDE

STEM SUPERWOMEN

Jane Goodall

By Amy B. Rogers

Cavendish Square

Published in 2026 by Cavendish Square Publishing, LLC
2544 Clinton Street Buffalo, NY 14224

Website: cavendishsq.com

Library of Congress Cataloging-in-Publication Data

Names: Rogers, Amy B., author.
Title: Jane Goodall / Amy B. Rogers.
Description: Buffalo, New York : Cavendish Square Publishing, [2026] | Series: The inside guide: STEM superwomen | Includes bibliographical references and index.
Identifiers: LCCN 2024054940 | ISBN 9781502674593 (library binding) | ISBN 9781502674586 (paperback) | ISBN 9781502674609 (ebook)
Subjects: LCSH: Goodall, Jane, 1934—Juvenile literature. | Primatologists–England–Biography–Juvenile literature. | Women primatologists–England–Biography–Juvenile literature. | Chimpanzees–Tanzania–Gombe Stream National Park–Juvenile literature.
Classification: LCC QL31.G58 R64 2026 | DDC 590.92 [B]–dc23/eng/20250218
LC record available at https://lccn.loc.gov/2024054940

Editor: Katie Kawa
Designer: Deanna Lepovich

The photographs in this book are used by permission and through the courtesy of: Cover, pp. 4, 16, 19, 24 vitrolphoto/Shutterstock.com; p. 6 Everett Collection/Shutterstock.com; p. 7 Nick Greaves/Shutterstock.com; p. 8 Melih Cevdet Teksen/Shutterstock.com; p. 9 Tinseltown/Shutterstock.com; p. 10 Carine Boekee/Shutterstock.com; p. 12 MehmetO/Shutterstock.com; p. 13 erwinf./Shutterstock.com; p. 14 Catherine D. Sales/Shutterstock.com; p. 15 Ferenc Szelepcsenyi/Shutterstock.com; p. 18 Helen J Davies/Shutterstock.com; p. 20 TetiBond/Shutterstock.com; p. 21 rustycanuck/Shutterstock.com; p. 22 Eugene Powers/Shutterstock.com; p. 25 James Kirkikis/Shutterstock.com; p. 26 1000 Words/Shutterstock.com; p. 27 spatuletail/Shutterstock.com; p. 28 (top left) max dallocco/Shutterstock.com; p. 28 (top right) Konstantin Yolshin/Shutterstock.com; p. 28 (bottom left) Kelleher Photography/Shutterstock.com; p. 28 (bottom right) TZIDO SUN/Shutterstock.com; p. 29 (left) Sean Fleming/Shutterstock.com; p. 29 (right) Dmytro Zinkevych/Shutterstock.com.

CPSIA compliance information: Batch #CSCSQ26: For further information contact Cavendish Square Publishing LLC at 1-877-980-4450.

Printed in the United States of America

CONTENTS

Jane Goodall turned her love of animals into a career as a scientist.

A LIFELONG ANIMAL LOVER

Chapter One

Many people around the world love animals, and some people turn that love into a career. Jane Goodall is one of those people. She's spent more than 60 years working with animals—from studying them in the wild to fighting to protect their natural **habitats**.

Jane's work as a scientist has changed the way we think about animals, especially chimpanzees, which she's been studying since 1960. However, she didn't start out with a plan to become an ethologist. She was just a little girl who was curious about nature—like so many kids are.

A Wartime Childhood

Jane Goodall has traveled and worked all over the world, but her story began in England. She was born there on April 3, 1934, to Mortimer Herbert Morris-Goodall and Margaret Myfanwe Joseph. Later on, the family grew when Jane's sister, Judy Goodall, was born.

Fast Fact

An ethologist is a scientist who studies animal behavior.

When Jane was only five years old, World War II (1939-1945) broke out. Many British men went to fight for their country, including Jane's father. This left her mother to raise her two daughters on her own for a long time. Jane's mother was strong and independent, though. In fact, she had her own career as a writer.

Fast Fact

Jane Goodall's full name is Valerie Jane Morris-Goodall.

All About Animals

Jane's childhood was filled with animals—both real and pretend. She had a pet dog like

Jane grew up during one of the most destructive wars in world history, which can be seen in this photo. It wasn't an easy time to be a child.

Long before Jane became famous for studying real chimpanzees such as this one, she was carrying around her favorite stuffed chimpanzee toy.

so many kids do, but she also spent time around other animals, including cows, tortoises, and ponies. In fact, when she was very young, she brought a bunch of earthworms into bed with her! When her mother found out, she didn't yell at her. Instead, she helped Jane take them back to their garden to keep them alive.

Jane's parents encouraged her love of animals. Her father gave her a stuffed chimpanzee named Jubilee. Jane still has that toy today.

Dreaming of Travel

Jane loved to spend time with animals and play outside. She also loved to read. Some of her favorite stories were about Tarzan—a man raised by apes in Africa. She also loved stories about Doctor Dolittle—a doctor who can talk to animals in their own languages.

Although these stories weren't real, they made Jane dream of going on her own real adventures someday. Because many of these stories

THE COLONIZATION OF AFRICA

For hundreds of years, European powers, including Great Britain, interfered with the people of Africa. This included enslaving Africans, forcing them to labor for no money in horrible conditions an ocean away from their home. Then, once enslavement had become illegal in most European nations and the United States, the Scramble for Africa began.

The Scramble for Africa is the name for a period from the late 19th century to the early 20th century in which European powers colonized many parts of the African continent. Great Britain had many colonies across the continent. However, after World War II, many of those colonies began to gain independence.

In many parts of Africa, the effects of colonization are still felt in the languages that are spoken, the traditions that have been lost, and the economic problems that have caused widespread poverty.

took place in Africa, Jane began to dream of traveling there to be like her favorite characters.

Making Money

As Jane grew up, she still held on to her dream of traveling to Africa. She studied the skills needed to become a secretary and worked at a variety of jobs instead of going to college. She worked as a secretary, as a waitress, and for a movie production company. This allowed her to save money for her dream trip.

Then, in 1957, Jane's friend from school invited her to come to a part of Africa that is now the nation of Kenya. At the age of 23, Jane packed her bags and took the trip that would change her life.

Fast Fact

Jane's sister still lives in the house where they grew up. This is also where Jane spends time when she's not traveling.

Travel has played a big part in Jane's life. Even today, she's said she's away from home around 300 days a year!

Jane's work with chimpanzees in Africa made her a respected primatologist, which is a scientist who studies primates. These are a group of mammals that include chimpanzees (such as the one shown here), gorillas, and humans.

STUDYING CHIMPANZEES

Chapter Two

Jane had always loved animals. From pets and farm animals to earthworms, she cared for them and paid close attention to them. Then, when she finally made it to Africa, she met a new animal friend who would become very special to her: the chimpanzee.

An Important Meeting

After Jane arrived in Africa, she met a famous scientist named Louis Leakey. Louis's work in Africa involved the origins, or beginnings, of humanity. He believed that studying primates would allow us to better understand these origins, because we **evolved** from a common ancestor.

Louis was looking for someone to study chimpanzees, and he believed Jane was the right person for the job. She was interested in animals, she was patient, and she was calm. She could also bring new ideas and ways of doing things to a scientific field that needed them.

Fast Fact

Louis Leakey believed women were better than men at studying primates because they were more patient and were seen as less of a threat to the people who lived in the areas where they set up camp.

Going to Gombe

In 1960, Jane began her study of chimpanzees. She set up camp in what's now Gombe Stream National Park in Tanzania. She was joined by her mother because, at the time, a young woman couldn't be in the area without a chaperone, or someone to watch over her.

At first, Jane watched the chimpanzees from a distance. Over time, though, the animals began to trust her. This allowed her to get closer to them and form **relationships** with them. She even gave them names.

Louis and his wife, Mary Leakey, found many clues to the origins of humanity in the Olduvai Gorge in present-day Tanzania, which is shown here.

THE TRIMATES

Jane Goodall was the first woman Louis Leakey sent to study great apes in the wild, but she wasn't the last. In 1966, Dian Fossey began observing mountain gorillas in their natural habitat in today's Rwanda. She was murdered in 1985, most likely because of her work to end poaching, or the illegal killing of animals.

Louis also selected a woman to study orangutans on the island of Borneo in Southeast Asia. Biruté Mary Galdikas began studying these apes in the wild in 1971. Together, Jane, Dian, and Biruté are sometimes known as the "Trimates."

All three of the Trimates wrote books about their experiences. Dian's book, *Gorillas in the Mist*, was turned into a movie in 1988, which was three years after her death. Shown here is her grave in Rwanda.

This was a big change, because other scientists used numbers instead of names for the animals they studied.

Fast Fact

The first chimpanzee to trust Jane was a male she named David Greybeard.

What She Saw

Jane was treated as a friendly neighbor by the chimpanzees, which helped her learn many important things about how they live. One of the most important things she discovered was that chimpanzees make and use tools. Before Jane made this **observation**, it was thought that only humans could do that.

Jane also noted that chimpanzees form lasting family relationships and have their own individual personalities and emotions—much like humans. She even discovered that

Jane's observations showed the world that chimpanzees were more like people than anyone had imagined.

Fast Fact

Another important observation Jane made was that chimpanzees sometimes acted violently toward each other. This behavior has been described as its own kind of warfare.

chimpanzees eat meat in addition to plants. Before her discovery, people thought chimpanzees were herbivores that only ate plants.

Big Changes

Jane's research changed the way scientists understood chimpanzees, and she'd done it all without a college degree. However, that changed in 1966, when she finished studying ethology at the University of Cambridge and became Dr. Jane Goodall.

Jane had become a respected scientist and a doctor in a short amount of time. Soon, she would also become a mother. In 1967, she gave birth to her son, Hugo. Becoming a mother didn't stop Jane's career, however. In fact, she was just getting started.

In 1965, Jane set up the Gombe Stream Research Center as the main research station to study chimpanzees. Today, it's still considered one of the most important "living laboratories" for the study of primates.

Jane's love for all animals influences her work as a conservationist. A conservationist is someone who works to protect the natural world.

BECOMING A CONSERVATIONIST

Chapter Three

Jane's work with chimpanzees gave her a deeper understanding of how amazing animals are. The more she learned, the more she wanted to protect the creatures she cared for. This included protecting them from one of the biggest threats to nature: humans.

The Jane Goodall Institute

Jane's work in Africa provided a new path forward for the scientific study of wild animals. She treated the chimpanzees she observed with respect and paid attention to their feelings. This view of animals as our neighbors informed everything Jane did, including the creation of the Jane Goodall Institute (JGI) in 1977.

JGI was created to help protect chimpanzees, which are an **endangered species**. It does this through the **establishment** of refuges, or safe places for chimpanzees to live in the wild, as well as through the sharing of knowledge about these special animals and other efforts.

Fast Fact

The Jane Goodall Institute works with communities to find conservation strategies, or plans, that help people and animals. JGI calls this method "community-centered conservation."

Chimpanzees are still endangered today, and some of the biggest threats facing these animals are poaching and the illegal pet trade. JGI works to educate people about these threats and laws that are meant to stop these practices.

A New Calling

Jane's work as a primatologist often took her to conferences, or gatherings of scientists who give talks on different topics. At a conference in 1986, Jane learned about the dangers of deforestation, which is the practice of cutting down trees in large numbers. Not long after that conference, she saw the reality of deforestation for herself when she flew over areas of Africa where the practice was destroying the natural habitats of animals.

Jane knew she had to do something. She used her voice to raise awareness about the importance of conservation. She was now an activist in addition to a scientist.

ROOTS & SHOOTS

Jane believes in the power young people have to create change in their communities and in the world. This is the driving force behind Roots & Shoots, JGI's youth program.

Roots & Shoots began in 1991, after Jane met with 12 teenagers at her home in Tanzania. They talked to her about problems in their community, and she was impressed with their desire to work toward a better future. This led to the creation of a program that focuses on educating young people about conservation and what they can do to protect the environment, or natural world.

Today, Roots & Shoots works in almost 100 countries around the world, teaching young people about the many ways they can help protect the planet.

Animal Advocacy

Fast Fact

According to JGI, in the last century, the number of chimpanzees in the wild decreased from 1 million to around 300,000.

Jane used her knowledge and passion for animals to become a powerful **advocate** for them. Her scientific discoveries about chimpanzees helped people see them as the advanced, intelligent, and **compassionate** creatures she knew they were. This helped her gain support for her efforts to end what she saw as cruel treatment toward them.

When Jane first began her work, chimpanzees were often used in medical experiments. This led to illness and death for many of these creatures. Jane has spoken out against this practice throughout her career, and her advocacy has helped end the use of chimpanzees in medical research in countries such as the United States.

A Trusted Voice

Jane has always been able to make science personal for people. By naming chimpanzees instead of using numbers, she allowed the public to make a personal connection to the chimpanzees she observed instead of viewing them only as scientific subjects. That ability to speak plainly,

Jane has become an important figure in the fight for animal rights.

personally, and passionately about ethology and conservation has made Jane a trusted voice in her field for decades.

For example, Jane worked with the Walt Disney Company to make sure its Animal Kingdom **theme park** was designed in a way that was good for the animals that lived there. When Jane speaks about animals, people listen.

Fast Fact

In 2015, the National Institutes of Health—the U.S. agency in charge of medical research—stopped using chimpanzees in its experiments.

Because of Jane's help in the creation of Disney's Animal Kingdom theme park, a carving of David Greybeard was put on the Tree of Life at the center of the park.

Jane's long, successful career has inspired many women to study animal behavior.

Chapter Four

A STEM ICON

Jane Goodall has been working for more than 60 years, and in that time, she has become a leading voice on conservation, an ethology **icon**, and a **role model** for women with an interest in science, technology, engineering, and math (STEM).

Today, the number of women in STEM fields is growing, and Jane has played an important part in that. By rising to fame as a scientist, she helped women see that it's possible for them to be successful in the world of STEM. That kind of representation matters. If someone sees someone who looks like them doing great things, it's easier for them to believe they can do great things too.

Fast Fact

According to the U.S. government, the percentage of women working in STEM fields is growing at a faster rate than the percentage of men working in STEM fields. Between 2011 and 2021, the number of women with STEM jobs in the United States rose from 9.4 million to 12.3 million.

Jane in the Media

Many people were first introduced to Jane through the media, including books, movies, and TV appearances.

Fast Fact

JGI helped create a TV show for kids called *Jane* that began airing on Apple TV+ in 2023. It's meant to inspire kids to think about Jane's work with animals and the ways they can help the environment.

Jane has written many books about her work with chimpanzees, including *In the Shadow of Man*, which was published in 1971, and *The Chimpanzees of Gombe: Patterns of Behavior*, which was published in 1986.

Many movies have been made about Jane and her work. For example, *Jane* came out in 2017, and *Jane Goodall: Reasons for Hope* came out in 2023. Jane has also appeared on many TV shows over the years to talk about her work and the importance of conservation.

Honoring Jane

Jane's impact on the fields of ethology and conservation has been recognized in many ways by countries and groups around the world. For example, she's been honored by the

Jane knows that it's important to teach young people about animals and the environment. That's why she's also written many books for kids.

French, Japanese, and Tanzanian governments for her work as a scientist. She also won the Benjamin Franklin Medal in Life Science, which is an American prize given to scientists for excellent work.

In 2002, Jane was named a Messenger of Peace by the head of the United Nations (UN). This is an international group that works toward goals including peace and human rights.

Fast Fact

In Great Britain, important men become knights, and important women become dames. Jane was named a Dame of the British Empire in 2004.

Helping Women

When Jane was first starting her career, primatology was a field that was mostly made up of male scientists. Today, women make up around half of all primatologists, which is a very good percentage for women in a

Jane believes that supporting women is good for the planet. She's a role model for many women working in ethology today.

"A SPEECH FOR HISTORY"

In October 2024, Jane gave a speech for the United Nations Scientific, Educational, and Cultural Organization (UNESCO). This speech was meant to make people feel hopeful about our planet's future. In "A Speech for History," Jane spoke about her belief in the power young people have to make a difference. She said, "Let us get together, each do our best in whatever way we can. Make the world better for the children that are coming after us and help them to make the world better for themselves."

Jane works hard to educate young people around the world about what they can do to help protect the planet from dangers that include climate change.

STEM field. Many female primatologists working today have said they were inspired by Jane to follow this path into the world of STEM.

Helping women in all areas of life is important to Jane. Through JGI, she works to give women around the world **access** to better education and health care.

A New Path

Jane Goodall didn't do things the same old way, and the world is better for it. She brought a new perspective, or way of looking at things, to the study of animals. This led to new discoveries about chimpanzees, humans, and all living things.

Today, Jane is still providing new perspectives, and she encourages young people, especially young women, to do the same.

Jane once said, "Every individual has a role to play. Every individual makes a difference." Her story has proven that this is true.

Jane has left such a mark on the world that she's been honored with her own postage stamps!

TIMELINE

In Jane's Life	World Events
1934 Valerie Jane Morris-Goodall is born in England on April 3.	
	1939-1945 World War II is fought.
	1955 Disneyland opens in Anaheim, California.
1957 Jane travels to Africa for the first time.	
1960 Jane begins studying chimpanzees in the wild in Tanzania.	
	1969 Astronauts walk on the moon for the first time.
1977 The Jane Goodall Institute is founded.	
	1998 Google is created.
	2020 The COVID-19 pandemic spreads around the world, leading to the deaths of millions of people.
2024 Jane delivers "A Speech for History" in October.	

THINK ABOUT IT!

1. What do you think is the most important thing Jane taught us about chimpanzees?
2. Why do you think some scientists disagreed with Jane's practice of naming the animals she studied?
3. How is supporting women good for the planet as a whole?
4. What are some things you can do to help care for animals and protect the environment?

GLOSSARY

access: The ability to use or have something.

advocate: A person who argues for or supports a cause or policy.

agency: A government department that is responsible for a certain activity or area.

compassionate: Having the ability to show deep sorrow for another's pain, as well as a desire to ease that pain.

endangered species: A kind of animal that is in danger of dying out.

establishment: The act of setting up.

evolve: To go through the process of evolution, or a series of changes in living things over time as they grow into new species.

habitat: The natural home for plants, animals, and other living things.

icon: A person who is admired for having a lot of influence in a certain area.

observation: A record of a scientific fact discovered by closely watching something.

relationship: The way in which two or more things are connected.

role model: A person who others look up to.

theme park: An amusement park in which all the rides and other buildings are based on a central idea.

FIND OUT MORE

Books

Klepeis, Alicia. *Jane Goodall*. Minneapolis, MN: Kids Core, 2022.

MacCarald, Clara. *Jane Goodall: Groundbreaking Primatologist*. Minneapolis, MN: Jump! Inc., 2024.

Woolf, Alex. *Jane Goodall*. New York, NY: Franklin Watts, 2020.

Websites

Jane Goodall Institute
Janegoodall.org/
The official website of the Jane Goodall Institute offers visitors information about Jane and the work JGI is doing around the world.

***National Geographic Kids*: Dr. Jane Goodall Interview!**
www.natgeokids.com/uk/discover/animals/general-animals/jane-goodall-interview/
This interview with Jane Goodall features her answers to important questions about her life and work.

***TIME for Kids*: 8 Questions for Jane Goodall**
www.timeforkids.com/g56/eight-questions-for-jane-goodall-g5-8-plus/
Visit this website to read an interview with Jane Goodall conducted by a student reporter.

INDEX